Courtesy of the author

About the Author

Andre Jordan is an artist, a daydreamer, and a writer. He has a regular column on the BBC's disability website, Ouch!, and his blog, ABeautifulRevolution.com—which deals mostly with his struggle through depression—has received more than two million hits to date.

Heaven Knows I'm Miserable Now

HARPER PERENNIAL

NEW YORK • LONDON • TORONTO • SYDNEY • NEW DELHI • AUCKLAND

Heaven Knows
I'm Miserable Now
ANDRE JORDAN

HARPER PERENNIAL

Originally published, in a slightly different form, as *If You're Happy and You Know It* in Great Britain in 2007 by John Murray, an imprint of Hachette Livre UK.

FIRST U.S. EDITION

Library of Congress Cataloging-in-Publication Data is available upon request.

ISBN 978-0-06-154730-0

09 10 11 12 13 ov/rrd 10 9 8 7 6 5 4 3 2 1

for stacey

Heaven Knows

I'm Miserable Now

Everyone thinks Possibility Girl is possibly a genius. Any day now, they continually agree, Possibility Girl will make it big. Become a star. 'You won't forget us when you're famous, will you?' they always say, as Possibility Girl begins yet another amazing project.

The only person who doesn't believe in Possibility Girl's possible genius, is Possibility Girl herself. She thinks they're being too kind. She isn't gifted at all. She's a fake genius, bluffing her way through life. She is convinced the moment she tries to actually achieve her full potential, she will fail, fall flat on her face, and the people that once admired her from afar will admire her no more. And so Possibility Girl never actually achieves anything. She just sits on the edge of her possible glory and basks in the adulation of her potential.

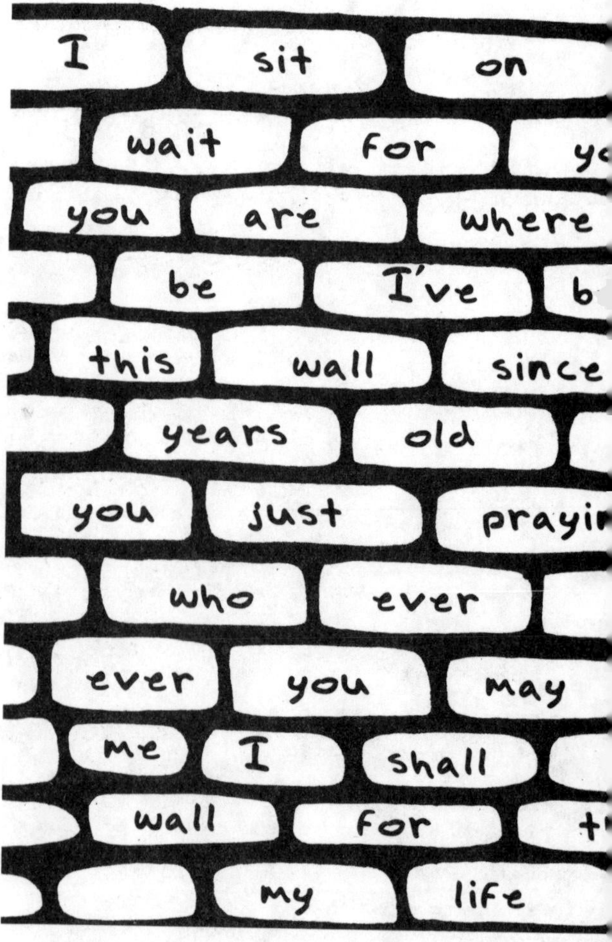

I sit on

wait for y

you are where

be I've b

this wall since

years old

you just prayi

who ever

ever you may

me I shall

wall for t

my life

his wall and

who ever

ever you may

sitting on

I was 17

t waiting for

for you

are where

be to find

on this

rest of

if need be

I
AM
DEPRESSED

please have sex with me

Girl with a borderline personality disorder

Me (I have run away)

You sit in the darkness. A double whisky in one hand, a cigarette in the other, and vow never to go speed dating ever again.

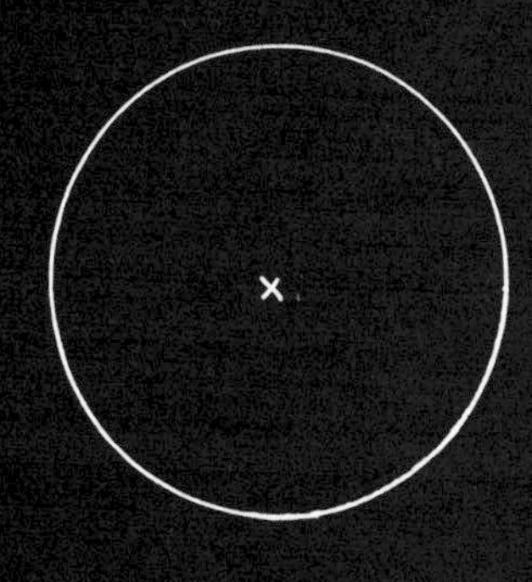

x: hate

z: everything

Gail 32. 5ft 6ins. small build
likes cinema. and the arts
seeks male. 30-40 for fun
and possible LTR

What Gail – likes cinema – forgot to mention were her scary psychopathic eyes and her dominatrix desires.

You have never eaten a pizza so fast in your life.

"Are we going to have sex now?"

"No, we are not."

It seems the whole world
is having sex apart from you

You go shopping for onions and love.
You have been told the best place to find
love is in the supermarket.
You wander the aisles for hours and hours
but to no avail.
You select two onions and go home and cry.

The Lovers just sit for hours
cooing at one another

It is disgusting!

QUANTUM PHYSICS

Oh how you hate
Friday fucking nights

LOVE

"There are plenty
more fish in the sea"

I
am
sorry

please
have sex
with me
again

Be still
my beating
heart

Goth girl – you make my heart twirl
You make everything – moody

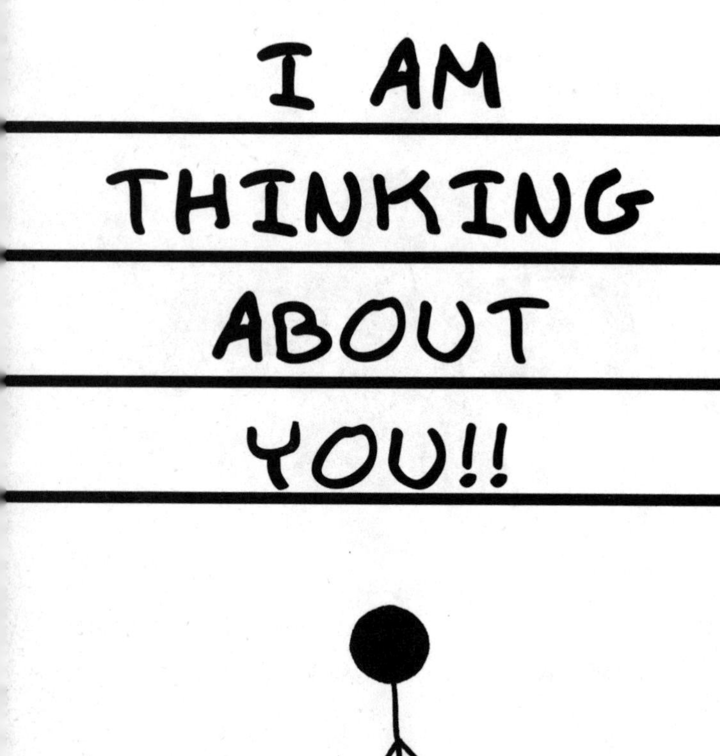
I AM
THINKING
ABOUT
YOU!!

this is what
your silence
over the
weekend →
did to me

I wrote a thousand love songs
and sang them from my heart.
But when I said they're all for you
you laughed and laughed and laughed.

At night I dream about
you a lot

Have you forgot about me?

Do I still give you
the ebejeebies?

It seems she does not love me.

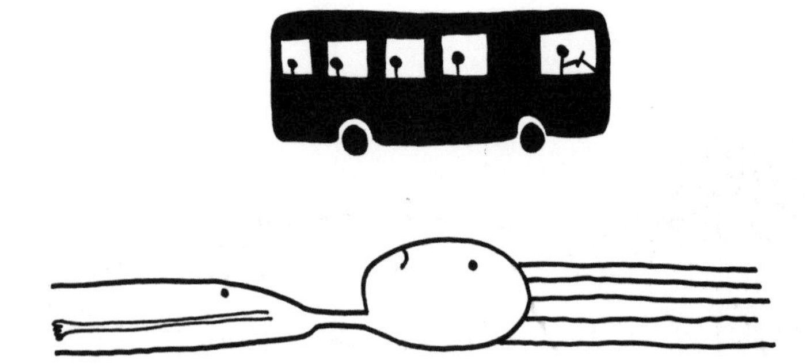

I shall lie in the middle of the road outside her house and wait for a bus to end this unbearable pain I feel in my heart.

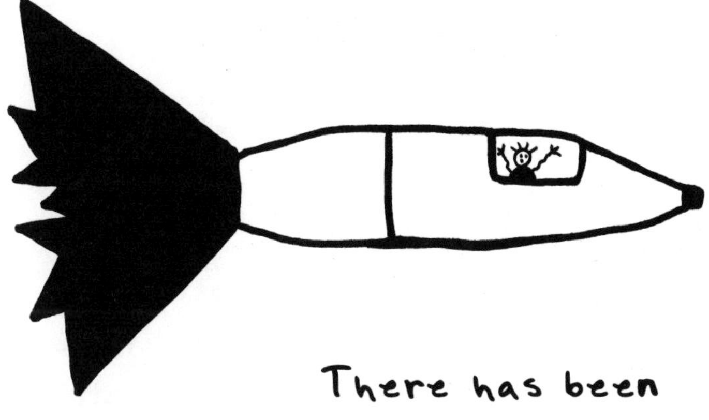

There has been
a terrible mistake

Your offer of love is declined.
She has chosen another.
You spend the rest of your days
painting pictures of despair.

Oh how her heartless message
has scared your delicate heart

The love-struck unfortunate
spends his days sniffing glue.

For he cannot bear to think of
another boy holding hands with you.

When you are lonely
life goes on and on
and on and on and on

Close my eyes and count back
from 10 9 been waiting all my life
been waiting all my life 8 7 it will
happen it will happen 6 5 seconds
left I am completely breathless
4 not sure not sure loves me
loves me not 3 don't go don't go
2 seconds remain nothing
ventured nothing gained
please stay...

I think about you
every day

They say 'Get a life'

I say 'NO'

My girlfriend

(please stop laughing)

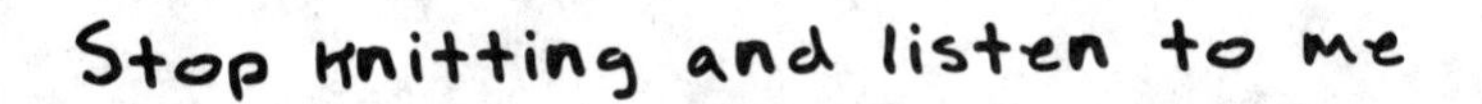
Stop knitting and listen to me

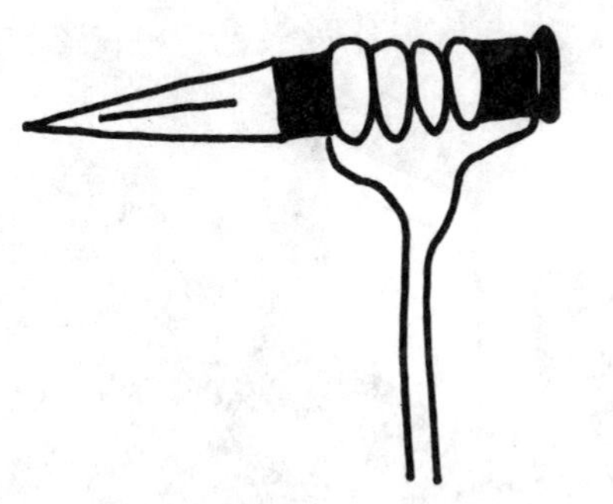

I retreat to the
garden shed
and wait for her
mood to lighten.

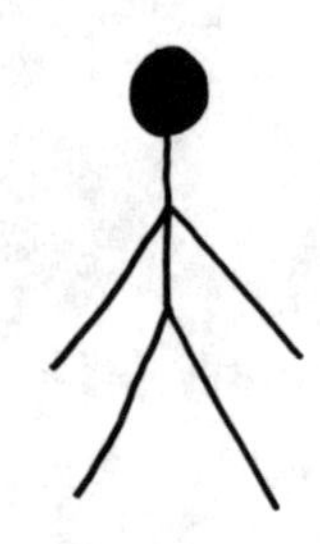

I have given you everything
but still you refuse to speak to me.

You are a rubbish girlfriend
and my mother was right about you.

Stick pin in appropriate place

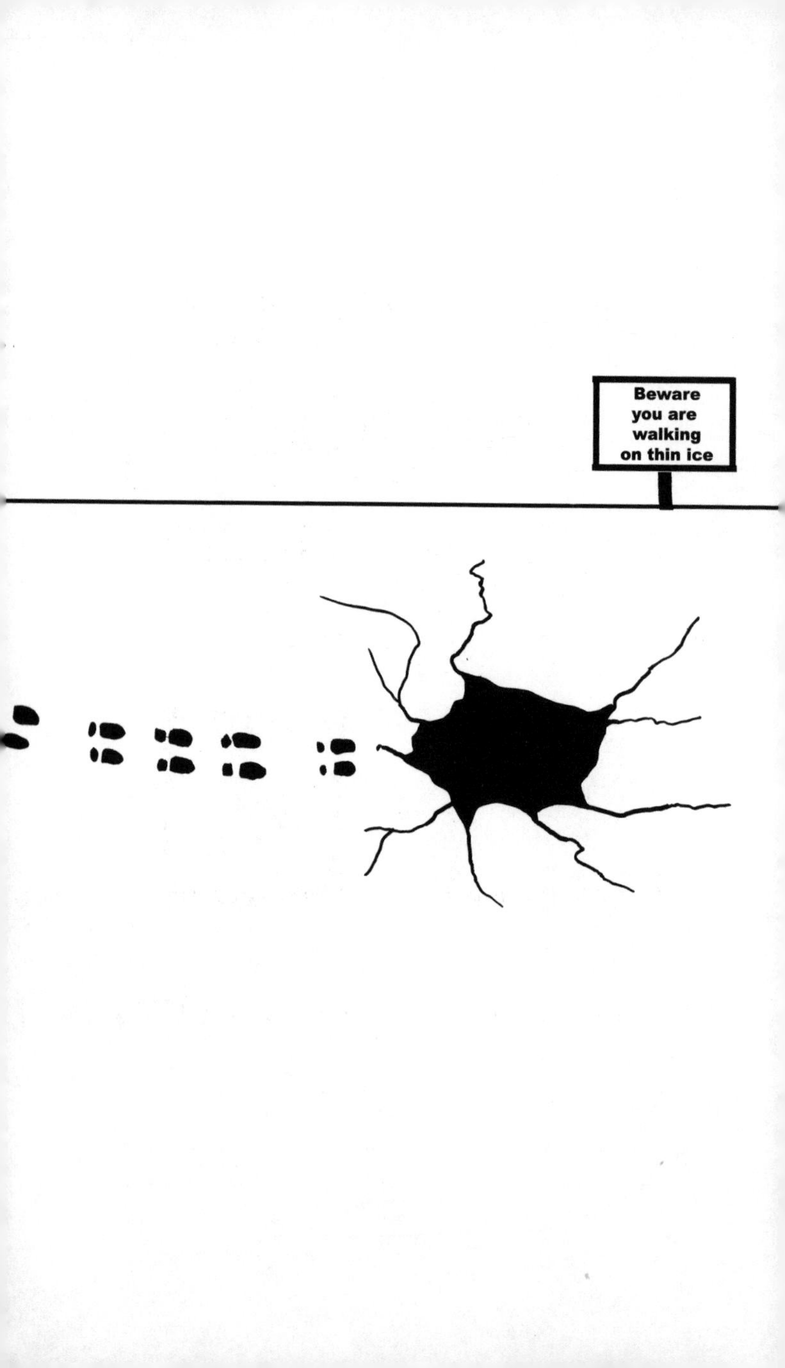
Beware
you are
walking
on thin ice

AM I STILL IN THE DOGHOUSE?

(please tick appropriate box)

"You're too intense"

Stop being so dramatic
I only said 'I might'
be pregnant

me

her

her father

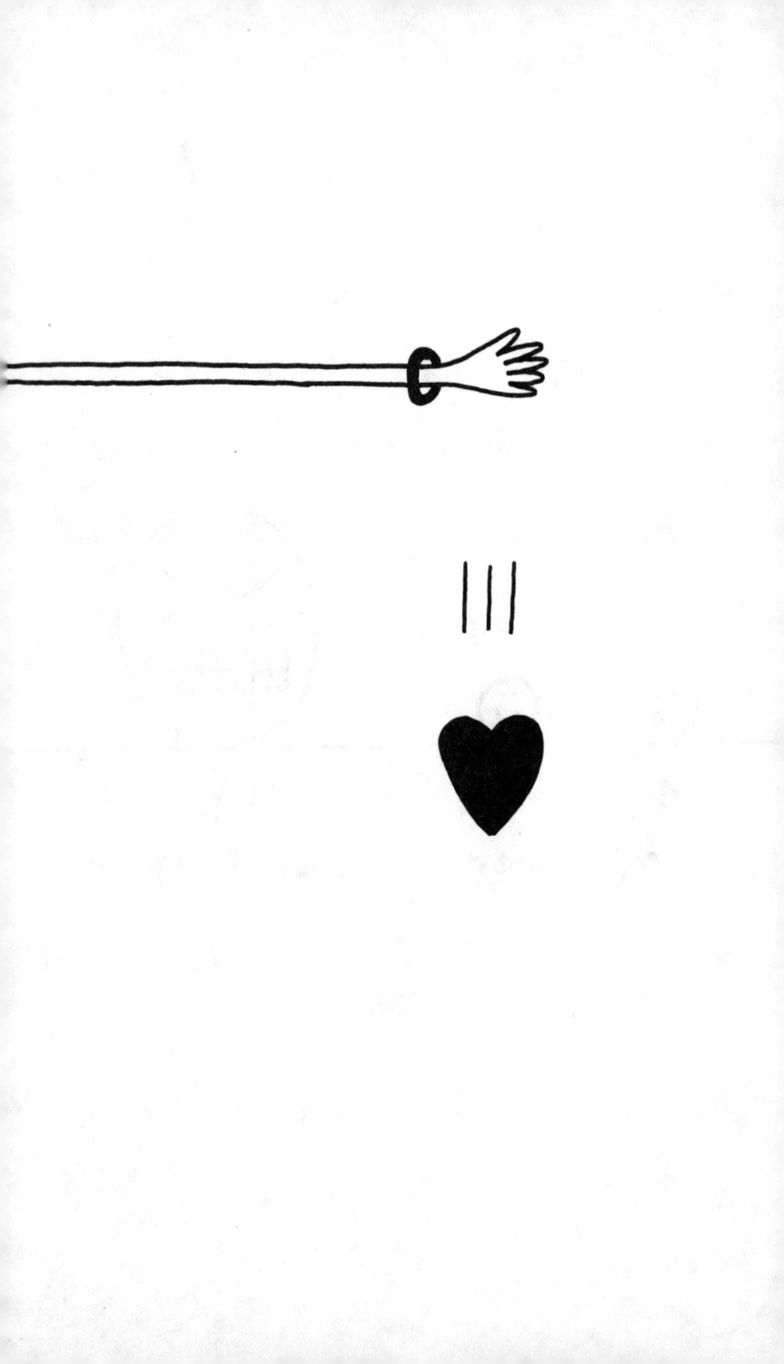

You and me

sitting in a tree

f i g h t i n g

I would like to kiss you

but you are sleeping
with that boy

I don't do triangles

No way, Jose

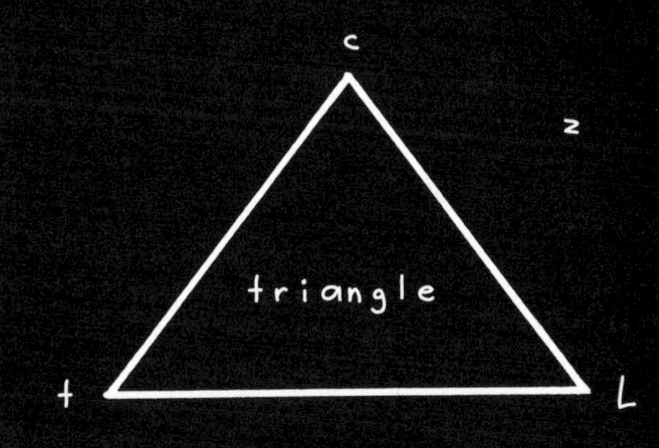

L: do not look at

t: do not think about

c: do not colour in

z = madness

PLEASE DON'T GO

I HATE YOU

She has left me
for a 'sunnier place'

My ex-girlfriend
(cow face)

She has gone.
All that remains is the faint
whiff of counterfeit perfume
and her beautiful toothbrush.

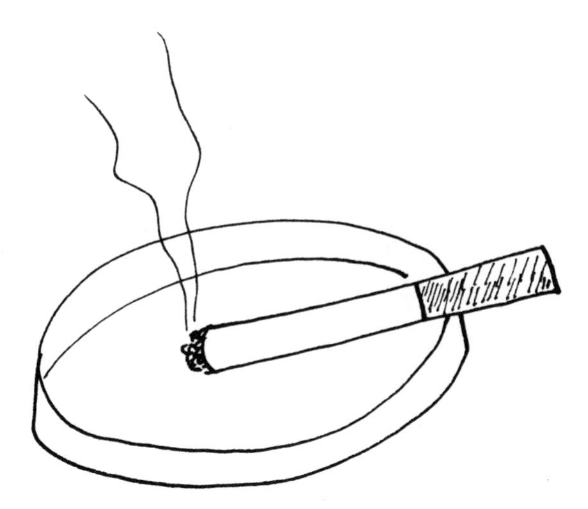

I miss you

I can't listen to this anymore

I have no idea how it feels
to be utterly loved.

I am the place a person falls
to when life gets hard. I am
the shoulder, the keeper of
secrets, the kindness through
their pain. I am the wisdom,
the knowledge, the prophet
when everything goes wrong.

I have no idea how it feels
to be utterly loved.

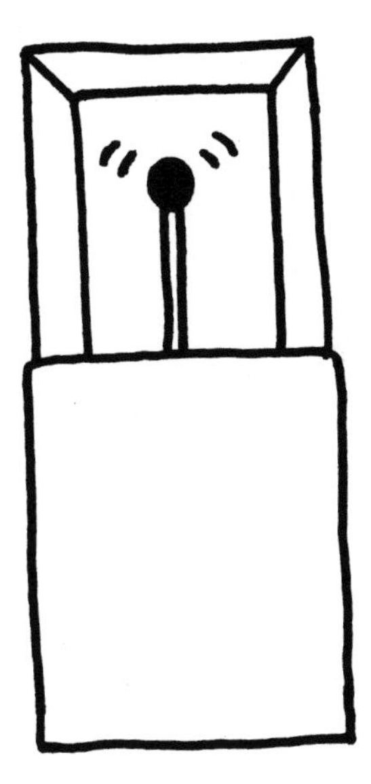

The matchstick's luck
had finally run out

everything is just rubbish

Take it, Dear Maker,
for I don't want it
anymore.

You are finally alone:
Masturbate

My only friend

still searching

I can't do this
send help soon
writes a boy in
a spacesuit
staring at the moon

Admit it
you are completely lost

THE BALL LAY IN THE GUTTER AND WAITED FOR SOMEONE TO PICK HIM UP AND PLAY WITH HIM. DAYS AND DAYS WENT BY BUT NO ONE, NOT EVEN THE LITTLE STRANGE BOY PICKED HIM UP TO PLAY WITH. IT WAS LIKE HE WAS CURSED. THE MOST DISGUSTING BALL EVER MADE. WHEN THE TOY FACTORY FIRST MADE HIM ALL THOSE YEARS AGO THEY MUST HAVE DONE SOMETHING TO MAKE HIM SO UNLOVABLE. SO HORRID. BUT WHAT HAD THEY DONE. THE BALL JUST COULD NOT UNDERSTAND. HE WAS JUST A BALL. LIKE EVERY OTHER BALL. AND YET HERE HE WAS, LYING IN THE GUTTER WAITING FOR SOMEONE TO PLAY WITH.

You only want to meet up when you are feeling depressed.

You never phone me when your life is good.

Look – do you want to go for a pizza or not?

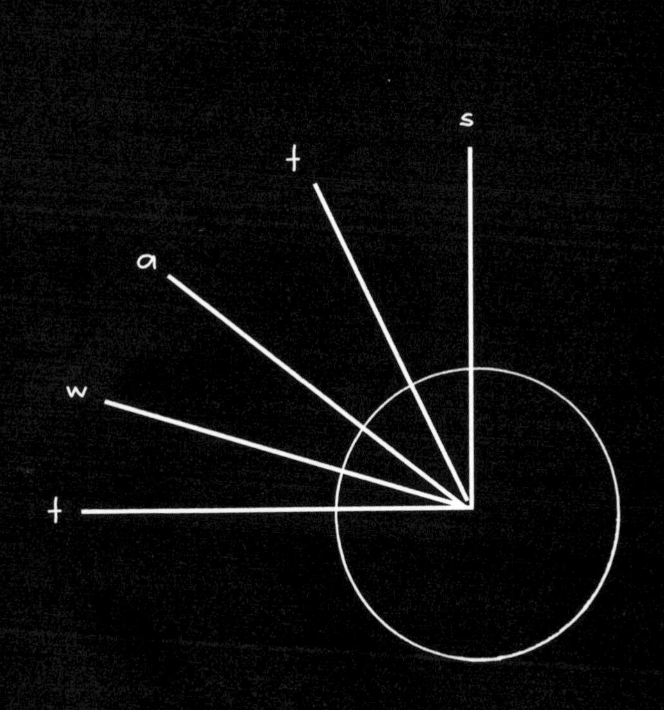

what most people are

A BAD NIGHT
at 3.07am the
white flag is
finally waved

Daytime
TV

The neighbours cat

I am bored
shitless

Je suis
un
idiot

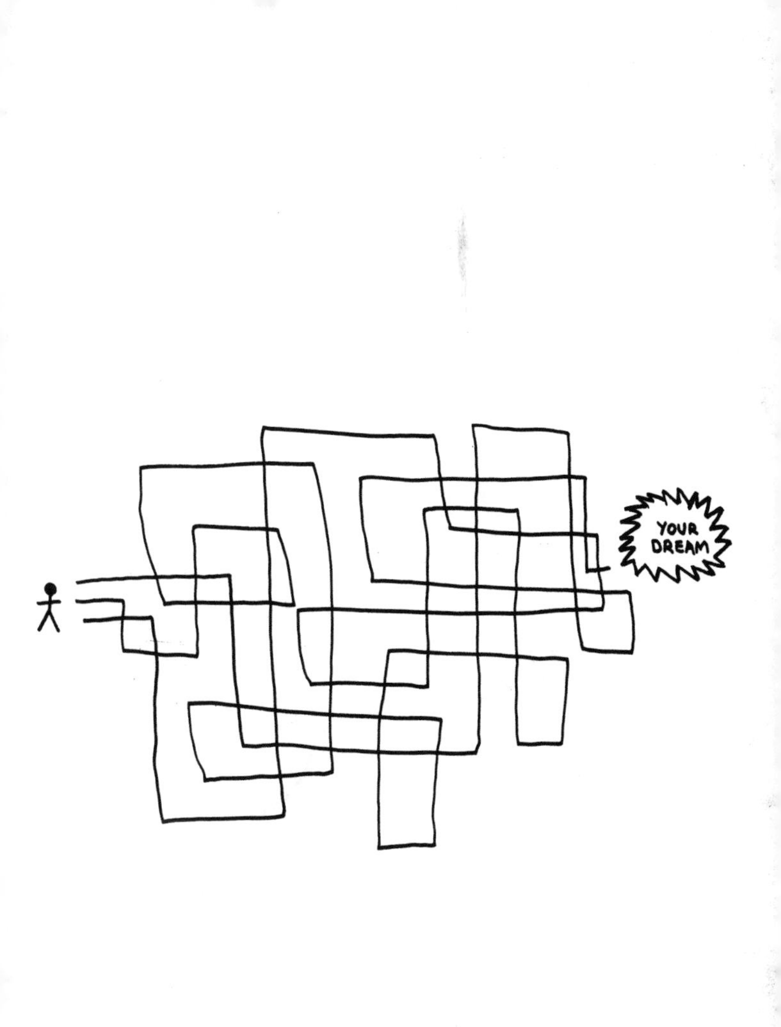
YOUR
DREAM

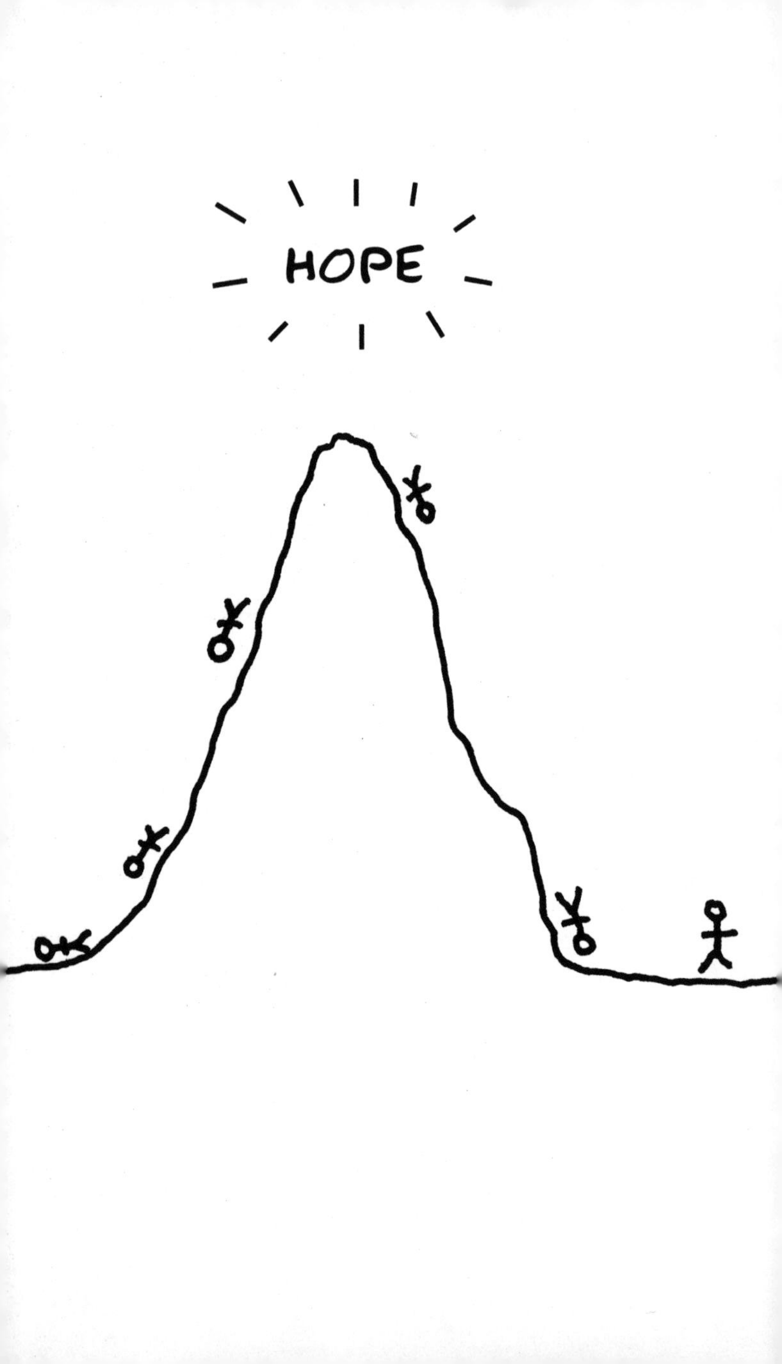
HOPE

Happy
Birthday

It is time to reflect upon your achievements thus far...

Oh how you cry and cry and cry

"You'll never amount to anything"

The saddest thing
that I ever saw
was once fearless
astronauts washed
up on earth's shore

Nobody shouts I'm
Spartacus anymore

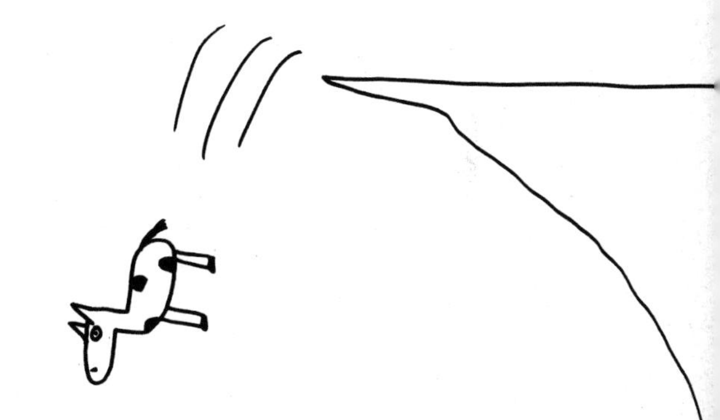

After yet another scathing review
the Pantomime Horse could take no more

They have died.
Every single one of them.
When Mother returns from
her holiday, she will kill you.

Lollipop shattered
on the floor.
Licked by worms
and slugs.
Everything ruined.
Complete disaster.

A Fatal Tale

'What are you doing?' asked the cat.

'I'm hiding,' said the partially blind mouse.

'What are you hiding from?' inquired the cat.

'Cats' whispered the mouse.

'But I'm a cat' purred the cat.

'FUCK !' yelped the mouse.

'It's okay' smiled the cat, 'I'm not very hungry at the moment.'

But it was too late – the partially blind mouse, presuming the worst, had thrown himself from the tree and was now lying dead on the pavement below.

MISFITS

Oh sweet baby Jesus make the bastard stop!

I fucking hate parties

The shy man shook with fear, as a small group of school children joined him at the bus stop.

I am a chicken

"Be a man"

CHILD
WRONG
PEACE
HELL
MONEY
RIGHT
ANGELS
SEX
DREAMS
GOLD
JOY
HOPE
FAME
HOME
EVIL
MOTHER
GOD

HOPES AND
DREAMS
PROHIBITED

The little book's glittering career
suddenly came to a humiliating end

Don't follow me
I haven't got a clue

Dear Grown ups,

I've been watching you for quite a while now and I have come to the conclusion that despite my age, I am not like any of you. I am not like any of you at all. Which is a shame really. And so I thought I would write to you in the hope that you and I might finally begin to understand each other a little better.

I think what happened was my brain stopped growing the moment I turned 14. To prove my point I have written a list of things I still don't understand.

Money
Mortgages
Business
Families

DIY
Insurance policies
Dinner parties
Man chats
Makeovers
Networking
Ambition

Anyway, I know you are always very, very busy and so I will keep today's letter brief.
But I will write again soon.

Sincerely yours,

Andre Jordan aged 41

Weirdo

Poor social
skills
have left me
isolated and
slightly
melancholy

- immigration control officer
- person wearing hugo boss
- person in newsagents
- person at bank
- person in post office queue
- her
- him
- you
- me
- hedgehog killed by car

I feel so utterly trapped

You travel this land
battered suitcase in hand
searching for someone
who will finally understand

Sometimes I just like to withdraw from the world and think about stuff.

(I am doing it now)

"Stop daydreaming"

Escapism No1

Escapism No2

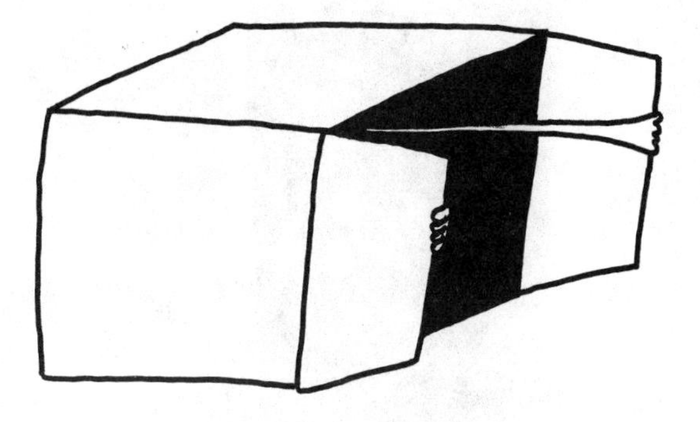

THE DOOMSDAY CODE

WƎ ARƎ FUCꓘƎD

In the dead of night
you bury all of your hopes and dreams

the dark place

i sometimes

go

TICK TOCK TICK TOCK TICK
TOCK TICK TOCK TICK TOCK

the rain

falls down

z: joy

y: adventure

a: german porn

Sperm comes from men
They make it in their sheds

I've decided to go
(it's for

and live in the woods

the best)

The chair waits
patiently for
you to sit with
him again.

You used to
talk about
everything

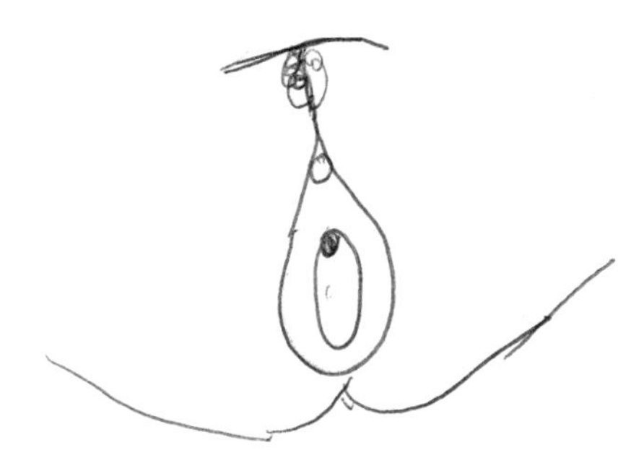

In the days
before internet
chat.

But now you
seem distant,
distracted.

And the chair
doesn't know
what to do.

You tell the
chair that
everything is
fine.

He is
worrying
too much.

It's just that your life has been a bit chaotic.

And a few things
have been playing
on your mind.

The chair
creaks slightly.

And waits
patiently for
you to sit with
him again.

I
HAVE TRIED TO
BE STRONG
BUT THE WEIGHT
OF MY LIFE
IS ENORMOUS

DEPRESSION

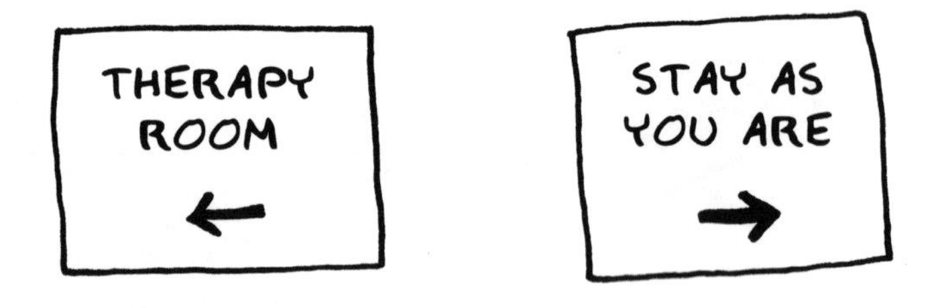
THERAPY ROOM
STAY AS YOU ARE

Trust me, I'm a psychotherapist.

Oh my God
look at that
It's unbelievable

HANDLE WITH CARE

"Tell me about your childhood"

My father once said
I was a
waste of space

I am determined
to prove him
right

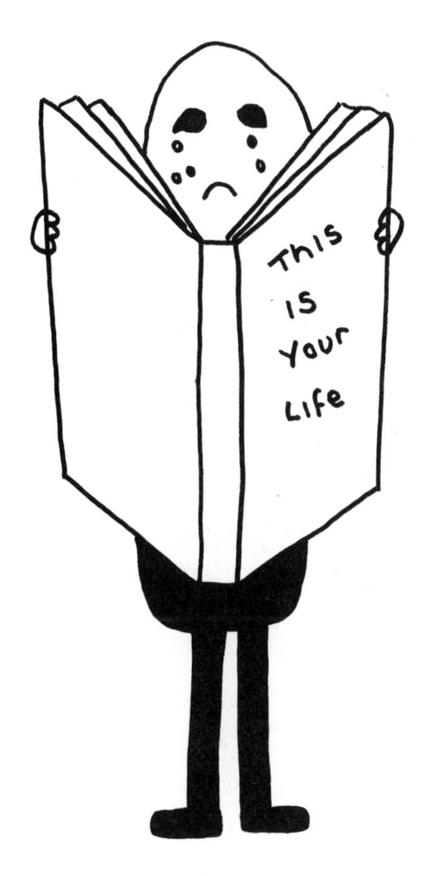
This
is
Your
Life

The place where your parents kept you captive for 18 years

Just thinking about it makes you shudder

I have
decided to
leave you
Earth Mother
I need to be
with people of
my own kind

I'd like you to close your eyes and pretend to be a baby

Oh for God's sake!

Evil Child

I once lived in a cul-de-sac
It was truly awful

The ghastly knitted jumper your Bearded Granny made you every Christmas.

Oh how it made you itch but still you had to wear it.

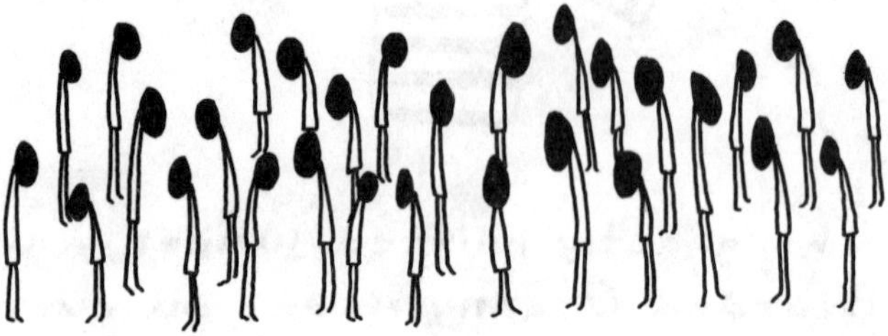

The School Reunion
(shudder)

Those dreadful disco nights.
Just thinking about the last
dance makes you want to cry.

Why did she reject you – why!

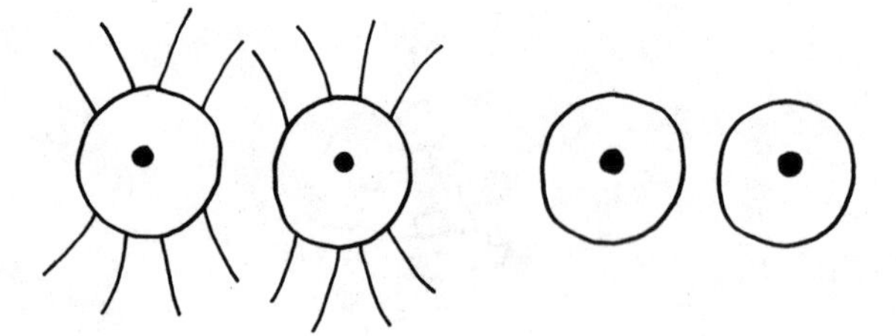

Your parents call a family meeting. They want to know what you are going to do with the rest of your life?

Despite being 40 - you still have no clue.

The shy sun never dared to shine

"And how did that make you feel?"

Oh how the battered umbrella
longed for the rain to stop

100 tears

It's hard to say when the depression began. It just appeared one winter's morning like an unexpected lunatic boarding your bus. And though you held your breath and tried not to stare as the lunatic made his way down the aisle with his carrier bags and Sony Walkman on full flipping blare, deep in your heart you just knew, you just knew – a matter of course that he'd sit next to you. And so you hold on tightly and gasp for air, as you take the ride of your life with a grinning lunatic on the seat next to you.

The hostage sees no way out

HELP

He's been doing
that for days

twat

I live in a box. It is warm and safe and dull. I would like to leave my box and go disco dancing with you but I am too afraid. I would like to leave my box and spend the rest of my life with you in the real world. But I am too afraid to do anything. That is why I live in this box. I hate this box.

Mirror mirror on the wall
how can you be so fucking cruel?

self harm

Self harm

'I hope it stops soon' said the girl

'Me too' said the boy

'I'm worried we'll get washed away' sighed the girl

The boy nodded, took the girl by the hand, and prayed with all his heart for the night storm to end

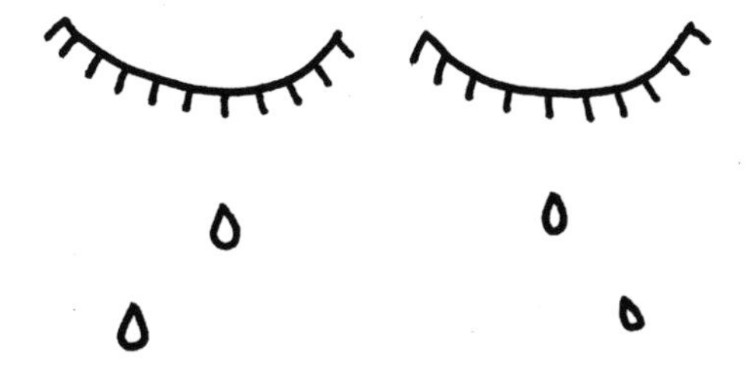

Sometimes life
just brings you down

I am not waiting for
something bad to happen

I am waiting for something
good to happen

"Pull yourself together"

Other
people's
expectations

I decide to hide in my favourite hole
and wait for the drama to pass

BIT MENTAL
PARANOID
SHY
ARTISTS

I am a
weirdo

Do not look at the elephant.

Pretend it isn't there.

Be Brave

Leave your pride at the door and just tell the truth

I am the Messiah

No, you are Bipolar

Hard
to
swallow

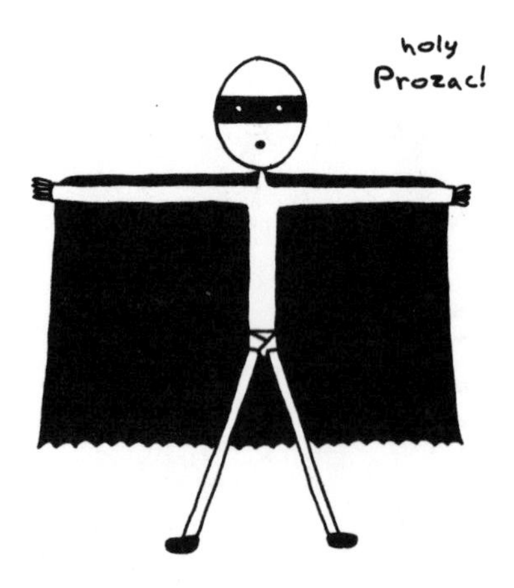

The Caped Crusader Of Gloom
has come to save us

I am diagnosed as being
schizoid passive aggressive

I quite like it

stop the war
No to nuclear
I am depressed
Please save me
Free Palestine

After two years of solitude,
I emerge from the manhole
practically unscathed.

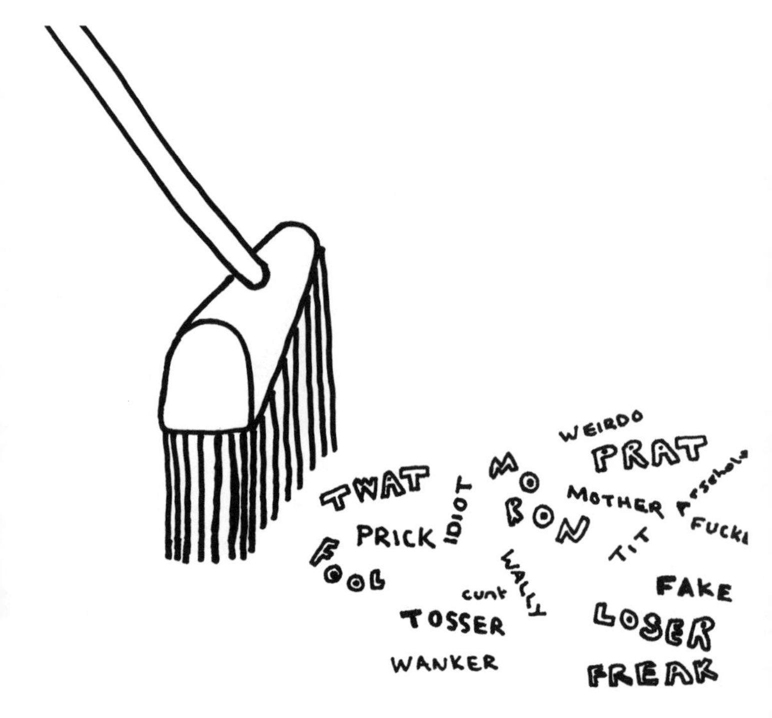
WEIRDO
PRAT
TWAT
IDIOT
MO
RON
MOTHER
PRICK
TIT
FOOL
WALLY
cunt
FAKE
TOSSER
LOSER
WANKER
FREAK

I am
MEEK
not
WEAK

Sex is good
Be Brave - Don't jump
FUCK
ME

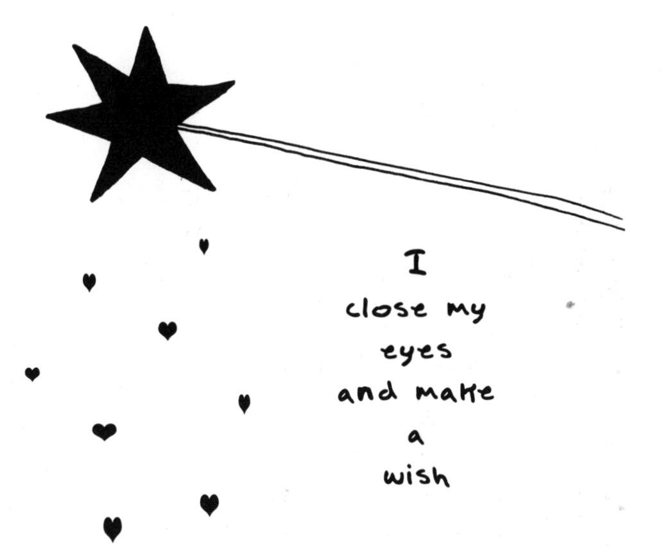
I
close my
eyes
and make
a
wish

Fear-less

'I'm leaving' said The Fear, dragging his suitcase into the hallway.

'Does this mean I shall be fear-less?' I asked.

'Not quite' said The Fear, 'Apprehension has decided to stay'.

I smiled, wished The Fear well and watched him nervously open my front door.

'Maybe see you again' he almost smiled, as he headed for the waiting taxi.

'Maybe' I said, not wishing to be rude. And then he was gone. I closed the door, took a deep breath, and smiled [apprehensively].

I do not wish
to alarm you
BUT I
once had mental
health issues

arse

DO NOT LISTEN

TO THOSE PEOPLE

WHO SAY I AM A

WASTE OF SPACE

THEY ARE WRONG

I AM ACE!!!

"Stop this creative nonsense and get a proper job!"

The key cost you £275,000
You wave it at everyone you meet
It is your greatest achievement in life
How proud you are of your £275,000 key

I have made some art. Do you like it?
I'm not sure. How much is it worth?

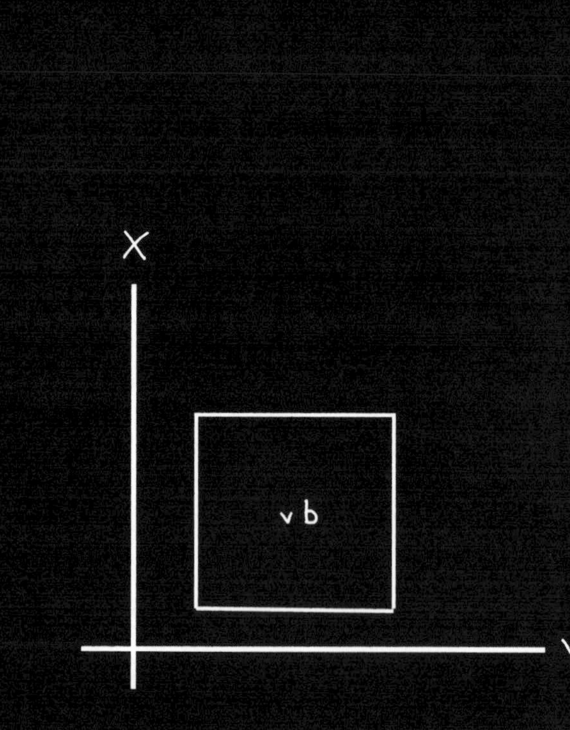

X: power games
Y: mind games
vb: very boring

you scare me

I fucking hate Christmas

Christmas isn't about you
It's about JESUS !!!

be nice

else you'll spend the rest of your life fighting and competing and you'll never trust anyone and you will end up just not being very happy and stuff

The early bird
ate too many worms
and died

Sticks and stones

are what

children throw

Don't let the ordinary change your extraordinary ways

I go out looking for astronauts. I walk slow. You might say I walk too slow. But someone once said they'd marry me because of my lack of pace. I walk even slower – just to make sure. Always the bridesmaid, never the bride. Always the bridesmaid, never the bride. I make a wish and watch it swim like a fish into the dog eat dog sky.

I walk as slowly as I can. Avoiding the cracks. Never looking back. It is my only plan. It is my only care. Please don't stare. I know they're out there. Astronauts. Hovering. Out of sync. Above the horizon. Below the radar. Graceful. Shy. Disconnected. Watching the world from way up high.

All my friends are astronauts.

By day he is a factory worker
But by night he is a Rock 'n' Roll star

"I've changed my mind – I want to live"

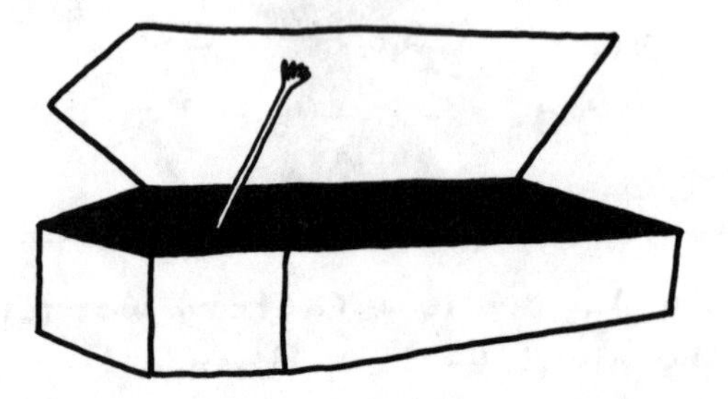

I'm too bloody happy

WELCOME
TO
HEAVEN

(please put down your guns)

The Clouds

I live in the clouds.
Reality is not for me.
People say I should come
down. That the clouds
are no place for grown ups
to be. I smile at them.
Maybe one day, I say, maybe
one day I will come down.
But I never will. Reality
is not for me. I shall stay
up here forever. The view
is quite breathtaking.

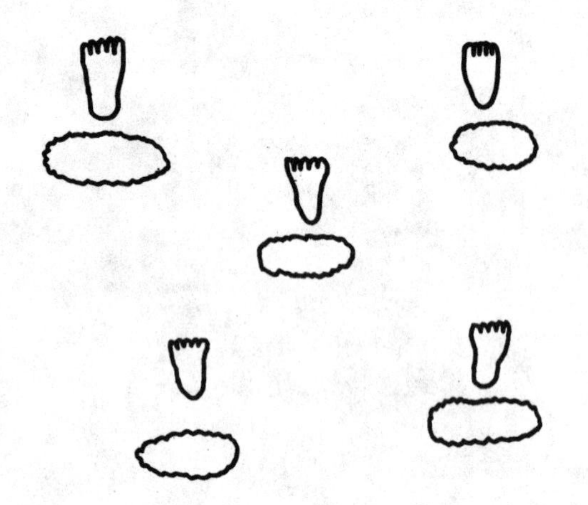

Only your soul goes to heaven. The rest of your body gets eaten by worms or becomes ash (depending on your personal preference).

Don't let go
(or look down)

You must never shout
at the concrete.

You must always shout
at the stars.

Concrete can't hear you.

I no longer
hear
the voices
in my
head

Every night I wrote a list of things that might go wrong.

Every morning I looked at the list and added a little more.

But then one day my pen ran out, and life completely changed.

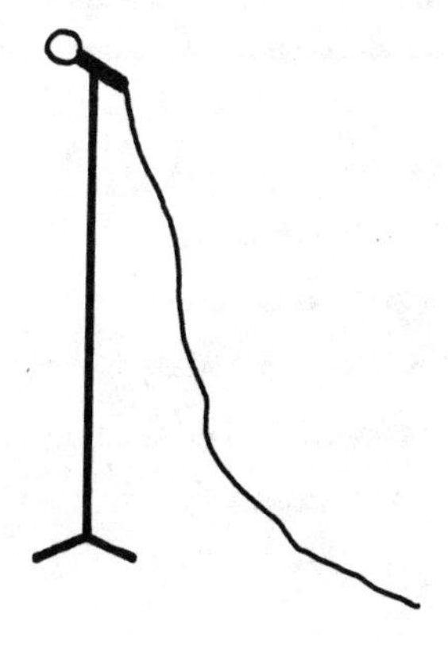

To achieve one's dreams
you must sing them – loud and clear
so that the Gods may hear

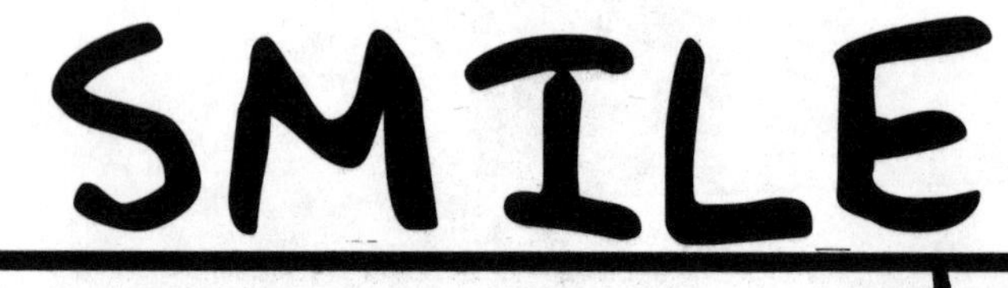

though your heart
is breaking and
you feel like a
misfit and no one
ever calls to ask
if you would like
to go to the
disco

There is a place that I once knew
Cold and frightening and bitterly blue
If you should find yourself there too
I'll hold your hand and walk with you

I wish I was sad once more. To see the girl that I adore. Those velvet lips I long to kiss. Oh how I miss my therapist.

HOPE
Don't
Let
Go

I do not care what car you drive where you live. If you know some one who knows someone who knows someone. If your clothes are this year's cutting edge. If your trust fund is unlimited. If you are A-list B-list or never heard of you list. I only care about the words that flutter from your mind. They are the only thing you truly own. The only thing I will remember you by. I will not fall in love with your bones and skin. I will not fall in love with the places you have been. I will not fall in love with anything but the words that flutter from your extraordinary mind.

All is not lost
all is never lost

Everything is possible

When you

strange

can be

are

world

amazing

love andre x

Acknowledgments

Thanks to Carrie, Brittany and all at Harper Perennial for their enthusiasm and guidance.